Changing Lives Through
3-D PRINTING

Other titles in *The Tech Effect* series include:

Changing Lives Through Artificial Intelligence

Changing Lives Through Genetic Engineering

Changing Lives Through Robotics

Changing Lives Through Self-Driving Cars

Changing Lives Through Virtual Reality

The Tech
EFFECT

Changing Lives Through
3-D PRINTING

Don Nardo

ReferencePoint Press
San Diego, CA

ReferencePoint Press®

For more information, contact:
ReferencePoint Press, Inc.
PO Box 27779
San Diego, CA 92198
www.ReferencePointPress.com

LIBRARY OF CONGRESS CATALOGING-IN-PUBLICATION DATA

Names: Nardo, Don, 1947- author.
Title: Changing lives through 3-D printing / by Don Nardo.
Description: San Diego, CA : Reference Point Press, Inc., 2021. | Series: Tech effect | Includes bibliographical references and index.
Identifiers: LCCN 2020013074 (print) | LCCN 2020013075 (ebook) | ISBN 9781682828472 (library binding) | ISBN 9781682828489 (ebook)
Subjects: LCSH: Three-dimensional printing--Juvenile literature.
Classification: LCC TS171.95 .N37 2021 (print) | LCC TS171.95 (ebook) | DDC 621.9/88--dc23
LC record available at https://lccn.loc.gov/2020013074
LC ebook record available at https://lccn.loc.gov/2020013075

CONTENTS

The Possibilities Are Endless

When asked how someone carves or creates an object from scratch, the gifted sixteenth-century Italian artist Michelangelo famously gave a sculptor's answer. Every block of stone has a specific statue hidden inside it, he explained, and it is the sculptor's job to reveal that statue to the world. In creating his great masterpieces such as the giant statue of the biblical figure David, Michelangelo worked from the outside in. Little by little, he carved away the layers of formless stone until David's magnificent physical form was released from the block that had imprisoned it.

Several centuries later, in the year 2019, scientists at the Swiss Federal Institute of Technology in Zurich, Switzerland, duplicated that feat by creating replicas of the statue. There were three noticeable differences between the old and new versions, however. First, the replicas are very tiny—almost microscopic. Second, they are composed of copper rather than marble. Third, and most importantly, the scientists reversed the original process by working from the inside out. Using a sophisticated 3-D printer, they fashioned the tiny metal Davids a little at a time by adding rather than removing layers.

3-D printing is a revolutionary, recently developed technique that allows people to make three-dimensional copies of objects far faster and cheaper than using traditional methods. The words on this page were created with a printer that discharges ink in a precise manner in two dimensions on a flat surface. That device can copy the same words over and over again as many times as a publisher desires. In contrast, a 3-D printer uses plastic, metal, or some other material, rather than ink, and in making its copies adds a third dimension—thickness. In this manner, the Swiss researchers made their little copies of Michelangelo's *David*.

Set to Transform Manufacturing

In doing so, the scientists say, they were demonstrating a printing technique that could have seemingly endless practical applications in the rapidly expanding field of miniature electronics. "Printing copper Davids is not the ultimate goal for anyone," remarks a spokesperson for the science news outlet Technology.org. "What scientists really want to do is to nail down the process to make other minute metal parts. 3D printing would be very helpful in, for example, manufacturing of some components for electronics. In fact, this technology could be used to repair some microelectronics as well. Or to print nanorobots. Possibilities are endless."[1]

nanorobots
Extremely tiny robots that can be injected into people's bodies

What is more, electronics is only one of hundreds, if not thousands, of manufacturing areas that 3-D printing is set to transform. For example, the companies that manufacture cars, airplanes, and ships are already experimenting with 3-D printing for fashioning the prototypes of their upcoming models. And using a larger 3-D printer, it is possible to reproduce exact copies of the actual full-sized cars—and in theory the planes and ships as well. In fact, 3-D printers show real promise in generating entire houses and most of their thousands of contents.

prototype
The first example or model of a newly introduced product

Inside a Typical Desktop 3-D Printer

Industrial-sized 3-D printers are still more common than desktop 3-D printers, but the latter have significantly improved over the last decade. The typical desktop device prints objects using a corn-based renewable plastic or other synthetic material **(1)**. To print objects, a robotic print head and extruder **(2)** operate like a hot-glue gun—layering the plastic on a build plate **(3)** that moves down as each layer of the object is printed. Using local controls **(4)**, the maker sets print settings that determine the object's size and other characteristics.

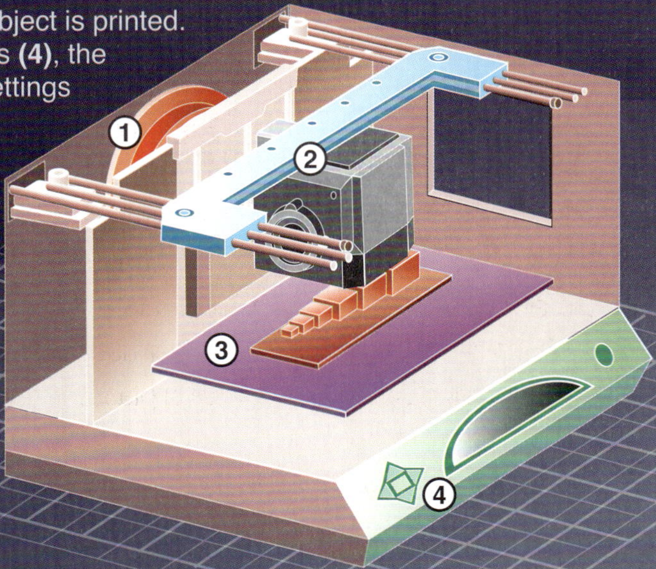

Desktop 3-D printers rely on a printing method known as fused deposition modeling, or FDM. To print an object using this method, the print head first makes an outline of the object on the surface of the build plate **(A)**. The outlined shape is then filled in with a cross-hatch pattern **(B)**. The build plate moves downward as new layers are outlined and filled **(C)**, eventually resulting in the completed object.

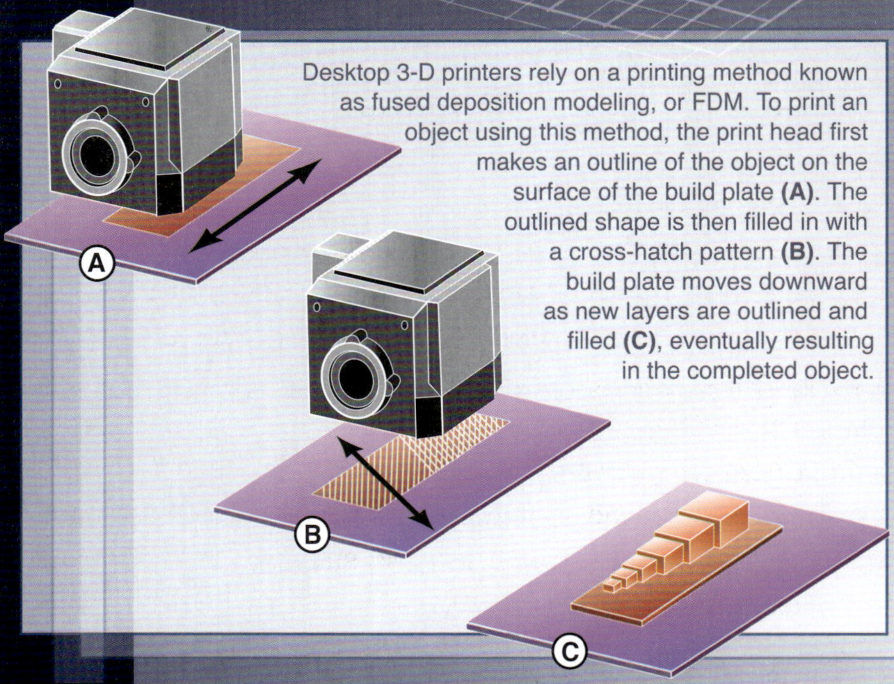

Source: *Wall Street Journal*, "MakerBot Replicator Mini Review: 3-D Printing Comes Home," June 17, 2014. www.wsj.com.

Creation by Adding Multiple Layers

The manner in which this happens is similar in some ways to how an ink-jet printer operates in conjunction with a computer. The ink-jet lays down a single layer of ink, re-creating the words and symbols that appear in a given computer file. In comparison, a 3-D printer lays down multiple layers, one at a time, creating three-dimensional thickness. In the course of only a few hours, the printer can create a three-dimensional model of fairly good size and complexity of detail.

One major difference between an ink-jet printer and a 3-D printer is the nature of the material each uses to do the printing. "Where an inkjet printer sprays liquid ink," British science writer Chris Woodford explains, a 3-D printer uses plastic. "A 3D printer essentially works by extruding molten plastic through a tiny nozzle that it moves around precisely under computer control. It prints one layer, waits for it to dry, and then prints the next layer on top. Depending on the quality of the printer, what you [can] get is . . . a stunning looking 3D model."[2]

Woodford adds that "the plastic from which models are printed is obviously hugely important."[3] In the world of 3-D printing, the word *plastic* does not necessarily refer to plastic in the usual sense of the term. Instead, the material merely has to display plastic, or highly flexible, qualities. In that regard, 3-D printers employ so-called thermoplastics, materials that melt when they are heated and turn back into a solid when they cool. Under that definition, most metals are thermoplastics and suitable for 3-D printing. Some researchers have even used chocolate because it takes on the forms of both a thick liquid and a candy bar–like solid. Depending on the particular design of the 3-D printer one uses, the plasticky substance can be fed into the printer in the form of either small pellets or thin strings that are rubbery to the touch.

By 2020 some of the more sophisticated 3-D printers were creating objects up to ten times faster than can be done using traditional methods and at a considerably lower cost. Although these devices are still expensive to purchase, in the area of tens of thousands of dollars even for basic versions, manufacturing experts are convinced they represent the wave of the future. As Woodford puts it:

> Many people believe 3D printing will herald not merely a tidal wave of brash, plastic gimmicks but a revolution in manufacturing industry and the world economy that it drives. . . . 3D printing will increase productivity (since fewer people will be needed to make the same things), lowering production costs overall, which should lead to lower prices and greater demand—and that's always a good thing.[4]

CHAPTER ONE

A Revolution in Transportation

"What made me flip and make the leap was this idea that plants and animals grow up very close to how 3D printers make things,"[5] says Canadian inventor and businessman Jim Kor. The creator of the world's first 3-D printed car—the Urbee—Kor observed that plants and animals grow up a little at a time, from one year to the next. A 3-D printer, he points out, creates an object little by little, one layer at a time, and in that way mimics nature's creation of living things.

Kor and his team deliberately made the Urbee as environmentally friendly as possible. Instead of running on gasoline or some other fossil fuel, which when burned pollutes the atmosphere, the Urbee uses electric motors, an energy source that Kor argues is more sustainable for human society in the future. "We should follow this direction if we want to be sustainable,"[6] he says.

According to Kor, the use of 3-D printers to make cars and countless other objects is rapidly expanding and will become the leading manufacturing technique of the future. Indeed, that bold prediction appears to be borne out by the facts. The International Data Corporation (IDC), an organization that regularly tracks consumer markets and services, found that in 2019 the global market for 3-D printing technology was valued at a minimum of $14 billion. That figure is expected to at least double in the next five years, the IDC states. Moreover, a large portion of that market has been,

One major market for 3-D printing is the manufacturing of parts for vehicles such as cars, airplanes, and even skateboards.

and is expected to remain in the near future, in the realm of transportation. Included are cars and trucks, spare parts for those vehicles, airplane parts, boats, bicycles, and even skateboards.

Automotive Uses

The continued expansion of 3-D printing in transportation will be especially evident in the Asian auto market, which experts say will grow by 30 percent or more each year at least through the mid-2020s. Driving that growth are the economies of existing industrialized Asian nations, including China and India, along with the economies of emerging nations in the region. All are exploring uses for 3-D printing in car manufacturing as a way to increase productivity and compete with their neighbors. In particular, China invested heavily in 3-D printing in its auto sector between 2010 and 2020 and will continue to do so in the next decade.

Meanwhile, some car companies in Western nations have also been investing large amounts of capital in 3-D printing. They hope to gain an edge on their growing list of competitors, both in Asia and the West. The Ford Motor Company is a good example. Already it frequently uses 3-D printing to make individual car and truck parts.

The original creator of the manufacturing assembly line, Ford found that 3-D printing technology could be smoothly integrated into its factory assembly lines. According to University of Mosul scholar Omar Jumaah, who has closely studied the recent growth of 3-D printing:

> Ford Motor Company has increased the total investment in its Kentucky Truck Plant to $925 million using 3D printing technology to increase the production of vehicles by 25%. Constructing a prototype part using traditional manufacturing processes may cost over $250,000 just in tooling, and take 8 to 16 weeks, while it can be done with a 3D printer in just hours or days, and cost only a few thousand dollars. As a result, 3D printing technology can make sense for production specific customized and complex casting applications with high quality.[7]

A Higher Volume of Mass Production

Older, larger auto manufacturers like Ford have not been the only big automotive investors in 3-D printing technology in recent years. From 2013 to 2016 Kor's company, Kor Ecologic, introduced the first two prototypes of the Urbee. Before that, no one had ever seen a car generated by 3-D printing technology, and to many people the idea had long seemed like something out of a science fiction novel.

Kor turned fiction into reality, however, and is marketing the Urbee with a global audience in mind. Because it is small as modern cars go and features three rather four wheels, the hope is that it will appeal to certain specialized or specific sectors of the population, especially in cities, where the Urbee will take up less space than traditional cars. As Kor and his team put it:

> For families we envisioned this as the second family car. For students, it could be their first car, and for retired people, perhaps their last car. Due to its long-life design, for some people it could be the only car they ever own. We believe Urbee is the perfect urban vehicle for the 21st Century. Every urban center in the world could benefit from adopting this [3-D printing] technology.[8]

A hybrid electric vehicle, the first version of the Urbee took twenty-five hundred hours to print. Since that time the company has managed to reduce the printing time for each vehicle by an estimated 30 to 40 percent. Some idea of printing times for such a vehicle can be seen in the case of the 3-D printer employed by the new California-based automotive company Hackrod, which aims to begin producing small customized cars in the early 2020s. Hackrod's 3-D printer is 6 feet (1.8 m) wide, 12 feet (3.7 m) long, and almost 4 feet (1.2 m) tall. It can print up to 5 pounds (2.3 kg) of aluminum per hour. At that rate, it will be capable of printing an entire small vehicle in only eight hundred to nine hundred hours.

hybrid
An object that combines two or more separate ideas or traits

To be competitive, of course, Hackrod, Kor, Ford, and other companies will need to have multiple printers of that type running simultaneously. Companies that make 3-D printers and sell them to the car companies have been rising to the challenge and making rapid improvements in those devices. In 2018, for example, Hewlett-Packard (HP) launched its Metal Jet printer,

The Urbee is the first car to be produced using 3-D printing technology.

which was specifically designed for high-volume manufacturing of production-grade metal parts.

In addition to employing such advanced printers, major auto companies realized that money and time could also be saved by printing the tools traditionally used by workers in automotive assembly lines. In the past, those tools had to be made separately, at considerable cost. By using 3-D printers to make both the auto parts and the assembly-line tools, these companies are on the verge of a new revolution in lower-cost manufacturing. Martin Goede, an executive in charge of technology development at Volkswagen, points out, "A single car consists of six thousand to eight thousand different parts." A big advantage of 3-D printing, he says, is that

> it allows us to produce many of these parts without first having to build manufacturing tools. By reducing the cycle time for the production of parts, we can realize a higher volume

of mass production very quickly. That's why HP's new Metal Jet platform is a huge leap forward for the industry, and we look forward to raising the bar on what is possible to deliver more value and innovation for our customers.[9]

Other Urban Vehicles

Small, fuel-efficient cars are not the only 3-D printed vehicles that transportation experts say will soon dominate travel within the world's larger urban areas. An Urbee-like car works well for a single person or a couple. But groups of four to fifteen or more people require means of navigating cities too.

For this challenge, the modest-sized Arizona-based automaker Local Motors has already provided a solution. Called the Olli, it is a 3-D printed, electric-powered minibus that seats twelve people. Small fleets of these innovative vehicles are already in the testing phase in some key US cities, including Washington, DC. They are needed, Local Motors chief executive officer (CEO) John B. Rogers points out, to fill the transportation niche of the so-called shared ride. "In the future," he says, "it is shared transportation that is organizationally owned [town buses and shuttles]. There will be shared transportation that is privately owned [taxi companies], and then there will be transportation that is not shared that is privately owned [family cars]. We'll have all these,"[10] Rogers states, adding that most of these vehicles will be made via 3-D printing.

organizationally owned
Owned and run by a company or business

Not all city dwellers drive cars or take buses or taxis, of course. Some ride bicycles and will continue to do so in the future. Hemant Bheda, chair of the California-based technology company Arevo, points out that several million bikes are manufactured each year using old-style industrial processes fueled by

Possible Downsides of 3-D Printing

Many scientists, engineers, and other experts have been touting the numerous upsides of the 3-D printing process, including in the area of transportation, for over a decade. Heard less often by the general public are a number of possible downsides for this new, revolutionary process. One, some experts point out, is that it requires a lot more energy to operate than most traditional manufacturing methods. A study by the research group Environmentally Benign Manufacturing, for instance, showed that in making a car, bike, or whatever, a typical 3-D printer used nearly one hundred times as much electrical energy as traditional methods. Advocates of the process say that new printers are more efficient and use less energy than earlier models, although the up-front energy needed to run these devices is still too high and needs to be reduced.

Another downside of 3-D printers is that unless they are very carefully constructed they can leak microscopic, possibly hazardous particles. Companies that make 3-D printers have acknowledged that this problem is real and have been actively working to eliminate this downside.

Still another disadvantage of 3-D printing involves counterfeiting and copyright infringements. "Anyone with a product blueprint can forge products very quickly," says Amada Pearson, a spokesperson for the online magazine 3D Insider. "Patent violations will increasingly become more common, and identifying counterfeited items will become practically impossible. As 3D printing technology evolves, patents and copyright holders will have a harder time protecting their rights and companies manufacturing unique products will be significantly affected."

Amada Pearson, "10 Disadvantages of 3D Printing Technology," 3D Insider, 2018. https://3dinsider.com.

burning coal or gas, both of which pollute the environment. This, he adds, makes bicycles a prime candidate for the much more environmentally friendly 3-D printing approach.

Bheda first saw the potential for that new 3-D printing transportation niche in 2016. Between then and 2018 he raised the millions of dollars in start-up money he needed to begin preliminary production of printed bikes. For a suitable design, he approached one of the leading bike designing firms, StudioWest, whose members,

The Olli is a 3-D printed electric minibus with seating for twelve passengers.

he recalls, "were excited by how the technology could push the limits of design."[11] All involved saw that printing an entire bike as a single unit would be far simpler than making it the traditional way, which requires assembling dozens of separate parts.

Bheda and the others decided to test the concept by printing a few prototypes. The StudioWest team provided a design and sent it to Bheda so his team could create the first model. "We then printed the bike," he recalls, after which "all of us rode it. It worked. It felt like a bike." Next, the StudioWest designers rode the bike. "They were extremely surprised that it rode exactly how they had specified it to be," Bheda says. "We wondered why they were surprised; they explained that they usually go through multiple iterations of design, ride, make some changes, ride again. It usually takes a long time, as much as 18 months, but in less than 14 days from the time that they sent the design, we had the bike."[12] All this convinced Bheda that he was on the right track. He gave the green light to production, and by early in 2020 Arevo was printing close to eight bicycles per week. Bheda has already ordered more printers and expects to eventually turn out several thousand bikes each year.

3-D Printed Aircraft?

Bheda also hopes that his company will soon begin work on printing more challenging vehicles, including what he calls flying cars. In this regard, he is not the first businessperson to consider the use of 3-D printing for the aircraft industry. A number of small aerospace and technology companies around the world are fast at work on prototypes for small 3-D printed hovercraft, often called "drone taxis." The companies say these will be used by police and public works departments, along with companies that deliver consumer products.

hovercraft
A flying machine that can remain stationary in the air above a single spot

Among the more promising examples is a flying car under development by Hoversurf, a company based in both San Francisco, California, and Moscow, Russia. By late 2020 or early 2021, the company plans to begin producing a craft that will carry five people and travel up to 185 mph (298 kph). Initially, a Hoversurf executive says, only individual parts of the craft will be created in a 3-D printer.

Some larger aircraft companies are also gearing up to make both new and replacement parts by printing them. The multinational aerospace giant Boeing, for instance, has begun making 3-D printed parts for jet engines. One of these is the GE9X, the sophisticated engine that powers Boeing's 777X, the world's largest twin-engine jetliner and passenger plane. All went as planned in a major test of those engines held in January 2020.

One obvious question that a number of scientists and nonscientists alike have already asked is whether it will ever be possible

3-D Printing a Boon to Car Collector Jay Leno

Although the day that average people can create entire cars in their garages using 3-D printers still lies a few years in the future, smaller, modified versions of that technology are already available. Comedian Jay Leno, former host of *The Tonight Show with Jay Leno*, has recently been in the forefront of this trend. Leno famously owns a large collection of classic cars that he keeps in a huge garage facility in Southern California. With the help of a handful of expert auto mechanics, he frequently involves himself in restoring various missing or damaged parts for his cars. When mass-manufactured parts are not available, especially for the older vehicles, Leno and his helpers turn to 3-D printing technology. They have made door handles, hood ornaments, and other parts this way, and the results have been spectacular. Leno said in a 2015 interview:

> Automobile manufacturing has changed more since 2000 to now than it did from 1900 to the year 2000. . . . This is pretty revolutionary. The 3D system is in the forefront [of new auto technology]. It's been an incredible boon to us that we can make any part for any car . . . [and do it] cheaper, faster, and more efficiently than they ever did. . . . This is the future, I think, of manufacturing in America.

Quoted in 3D Systems, *Jay Leno on 3D Scanning and 3D Printing from 3D Systems*, YouTube, December 8, 2015. www.youtube.com/watch?v=OzlGfclyBnc.

to go beyond merely 3-D printing aircraft parts and make an entire large-scale airplane using a 3-D printer. The answer appears to be yes. But inventors and other researchers must first overcome one fundamental obstacle: no 3-D printer large enough to make an airliner yet exists. One expert, Bastian Schaefer, an engineer at the European aircraft manufacturer Airbus, has considered the problem. He estimates such a printer would need to be at least 262 feet (80 m) long—almost the length of a football field. Yet he is confident such a device could be built using existing technology. "This could be feasible,"[13] he remarks.

The Biggest Printed Object Yet

Although no one has yet 3-D printed an entire airplane, another popular form of transportation—a seaworthy boat—*has* been created in a 3-D printer. This notable feat was accomplished in 2019 by a team of researchers at the University of Maine. By far the biggest single 3-D printed object yet created, it is 25 feet (7.6 m) long and weighs 2.2 tons (2 metric tons). Its makers christened it the *3Dirigo*.

Production of the *3Dirigo* required a very large, specialized printer. With the aid of a team from Ingersoll Machine Tools in Rockford, Illinois, a University of Maine team fashioned the one-of-a-kind device, which measures 98 feet (30 m) long, 21 feet (6.4 m) wide, and 10 feet (3 m) high. It was specifically designed to use raw materials of biological origin, especially cellulose, a tough fibrous material found in wood. The process of printing the boat took a little more than seventy hours.

For the university and Ingersoll teams, the project was not about a love for ships and the sea as much as it was about testing what will be possible for 3-D printing technology in the 2020s, 2030s, and beyond. "If you'd heard just 10 years ago that a 25-foot boat was going to be made by a printer, you'd never have believed it,"[14] science writer Marc Prosser said after examining the *3Dirigo* up close. But as he and other experts point out, 3-D printing is reaching much of its technological potential in the realm of transportation. There, as in some other manufacturing areas, that seemingly magical process is making it easier to create innovative products of all kinds faster and at lower cost.

An Entire House and Its Contents

During the summer of 2018, French couple Nordine and Nouria Ramdani and their three children received some singular publicity as newspapers, radio and TV talk shows, and online magazines across Europe and beyond interviewed them. Their sudden and temporary brush with fame did not stem from musical or other artistic talents. Nor did they say or do anything socially or politically outrageous. Rather, they made news because their family was the first ever to permanently inhabit a house created by 3-D printing.

The 1,022-square-foot (95 sq. m) home, located in Nantes, France, was developed by researchers at the University of Nantes. It took them fifty-four hours to print the structure, but very soon afterward improvements in the printer they had recently built reduced the time required to make their second house to just thirty-three hours. The cost of the Ramdanis' new home was $232,000, roughly 20 percent less than houses that size then cost to build in France using traditional construction methods.

The attractive price was certainly a factor in the Ramdanis' decision to buy the printed house. But there were other reasons, too, including the fact that they would be helping make history. "It is a great honor to be part of this project," Nordine said in a September 2018 interview. "We lived in a neighborhood of social housing dating from the 1960s. So this is a big change for us. Here we have a house and a garden. It's something else!"[15]

Using a 3-D printer to make an entire house is unarguably unusual in itself. Even more extraordinary, however, is the fact that in theory practically the entire contents of a typical modern house could be fashioned by such printers, either now or in the very near future. In the words of technology writers Adrian Willings and Luke Edwards:

> The world of 3D printing is well and truly here with printers in homes, online stores, and even supermarkets offering 3D printing services. The result is all manner of weird and wonderful 3D printed objects in the home. People can download or design files and print off objects in plastic. This can be helpful for cheap and easy replacements of parts about the house, quick toys, and even furniture.[16]

Willings and Edwards did not mention clothes, kitchen appliances, and food, all of which have already been 3-D printed.

The Affordability Factor

The University of Nantes 3-D printing team, headed by Professor Benoit Furet, approached the home-printing project literally from the ground up. Traditional home builders do the same thing, by first pouring a concrete foundation and then adding wooden walls and a roof on top of it. The French team's printer laid down the entire house, layer by layer, from the bottom up. Each wall featured two panels—an inner one and outer one—with a space a few inches wide between them. After the printing phase concluded, a crew poured concrete into that cavity, thereby producing a strong, well-insulated barrier to the outside elements.

In a 2018 interview with the BBC, Furet pointed out that as relatively inexpensive as the house was to create, later versions would be even cheaper to produce. This is because as time goes on, the 3-D printing process will become more advanced and streamlined, and the printers themselves will cost less to make

Using technology developed by a team of researchers from University of Nantes in France, a construction worker creates an entire house using a 3-D printer.

and buy. By 2023, after only five years, Furet predicted, the printing process will be at least 25 percent less expensive. Moreover, he said, just five years after that—in 2028—it will be up to 40 percent cheaper. So a printed house identical to the one the Ramdanis moved into could conceivably go for as little as $140,000 in 2028.

Other companies that have begun 3-D printing houses—of which at least two dozen existed worldwide in late 2019—have forecast similar decreases in start-up prices during the 2020s. One of the leaders in this market in the United States—Haus .me, based in Reno, Nevada—has already significantly reduced its overhead by building a central assembly plant in Reno. Specially designed flatbed trucks carry the finished homes to sites designated by buyers in Arizona, Nevada, and California.

These homes can go from the planning stage to completion in only a few weeks, compared to the months it typically takes to erect new houses the old-fashioned way. According to Haus. me CEO Max Gerbut, each printed house is self-contained. It has all the features one expects in a modern home, including running water, electricity, proper insulation, and even resistance to earthquakes. Yet as is the case with other companies that create printed houses, the price remains low.

exponentially
A lot, or to a great degree

This affordability factor has been the chief driver of the 3-D printed housing market, says Brett Hagler, director of New Story, a nonprofit group that seeks ways to end global homelessness. New Story has been working with one of Haus.me's competitors, the American construction firm ICON. "We feel it's our responsibility to challenge traditional methods and work toward ending homelessness," Hagler states. Traditional house-building methods, he argues, "will never reach the billion+ people who need safe homes. By working with ICON and leveraging their 3D printing innovations, we're able to reach more families with the best possible shelter solutions, exponentially faster."[17] Jason Ballard, one of ICON's founders, agrees. "Conventional construction methods," he says,

quantum leap
A tremendous, rapid advancement in a technology or an approach to doing something

> have many baked-in drawbacks and problems that we've taken for granted for so long that we forgot how to imagine any alternative. With 3D printing, you not only have a continuous thermal [heat-containing] envelope, high thermal mass, and near zero-waste, but you also have speed, a much broader design palette . . . and the possibility of a quantum leap in affordability. This isn't 10% better, it's 10 *times* better.[18]

Printed Chairs and Other Furniture

Thanks to the advent of 3-D printing technology, modern design palettes for the home now extend far beyond the design and construction of the house itself. All houses contain typical items, not least of which is an array of furniture. Realizing the potential of 3-D printing for their business, between 2010 and 2020 every major furniture maker in the world established at least one line of printed products.

Indeed, the world's largest furniture manufacturer, Wisconsin-based Ashley Furniture, has strongly embraced the new technology. The company produces more than three hundred thousand furniture items per week in factories located in Wisconsin, Mississippi, California, Indiana, Pennsylvania, North Carolina, and Florida, as well as abroad. Initially, Ashley installed two 3-D printers in its Wisconsin facility and immediately found a significant reduction in basic costs. It also noticed a new flexibility in both design and production. The printing process allowed for the creation of chairs, tables, and other items with shapes more complex and in-

Ashley Furniture uses 3-D printing to produce chairs, tables, and other items for use in the home.

Jobs Lost, Jobs Gained

As 3-D printers become more competitive with traditional manufacturing methods for items such as clothes, houses, cars, and more, workers who traditionally make these products could lose their jobs. "3D printing has the power to make manufacturing far less expensive and more automated," says technology reporter Frank Landman. "Accordingly, we may see the loss of some human jobs in the manufacturing industry" as 3-D printers overtake traditional methods of manufacturing.

Nevertheless, many experts on emerging technologies argue that in the long run 3-D printing will create more jobs than it eliminates. It will do this, not by replacing existing manufacturing industries, but by creating new ones. "Why does it have an advantage over other making technologies?" asks Kerry Stevenson, CEO of the popular 3-D printing news source Fabbaloo.

> It's because the nature of the making process allows for the creation of objects that could not be made in other ways. In other words, the technology can produce impossible parts. Or at least geometries that our long training in non-3D printing design has taught us could not be built. I think the key here is to leverage 3D printing to develop new, "impossible" parts that either revolutionize industries or create entire new industries. That is something that would create jobs.

Frank Landman, "How Will Home-Based 3D Printers Impact the Economy?," *ReadWrite* (blog), May 17, 2018. https://readwrite.com.

Kerry Stevenson, "How 3D Printing Can Really Bring Back Jobs," Fabbaloo, September 21, 2019. www.fabbaloo.com.

tricate than were possible using traditional construction methods. As Ashley production engineer Brian Konkel puts it, 3-D printing "gives us the ability to quickly adapt to any suggestions or to any needs that we have, and create geometries [shapes] that we're not traditionally able to manufacture or machine. We're able to do things that weren't quite possible several years ago."[19]

Other large furniture makers have adopted 3-D printing technology, among them the noted Swedish company IKEA, which has plants and stores around the globe. Starting in 2018 the company introduced a new line of printed products. Most can be used as add-ons to existing IKEA merchandise. In addition, the up-and-coming company Print the Future hopes to carve out its own major niche in the home furnishings market. "What Starbucks is to coffee," boasts company CEO Neil Patel, "Print the Future will be to 3-D printing, synonymous with its product."[20]

Some furniture companies have devised clever, as well as practical, ways to use 3-D printing in environmentally friendly ways. A noteworthy example is the Strat line of printed furniture made by Belgian designer Joachim Froment. All of the stools, chairs, tables, and other products he creates are made from plastic waste recycled by homes and businesses in his own region. "The aim of this project," Froment explains, is to offer "a service to recycle the products in order to close the loop and participate entirely in a circular economic system. This way of design together with the technology could compete with current industrial processes [and introduce] a potential new way to create sustainable objects at an affordable price."[21]

Kitchen Appliances and Food

In addition to movable furniture, every house features built-in appliances, the most visible and important being those in kitchens. Some companies are already manufacturing refrigerators, ovens, meat cookers, and other appliances using 3-D printing technology. The latter "has staked its claim as an effective and innovative tool for food processing," expert observer Charles Goulding points out, "with several companies taking advantage of the limitless benefits that integrating the technology can provide."[22]

At the forefront of those companies is Louisville, Kentucky–based FirstBuild, which makes state-of-the-art refrigerators. Its 3-D printed model the ChillHub, released in 2018–2019, is far

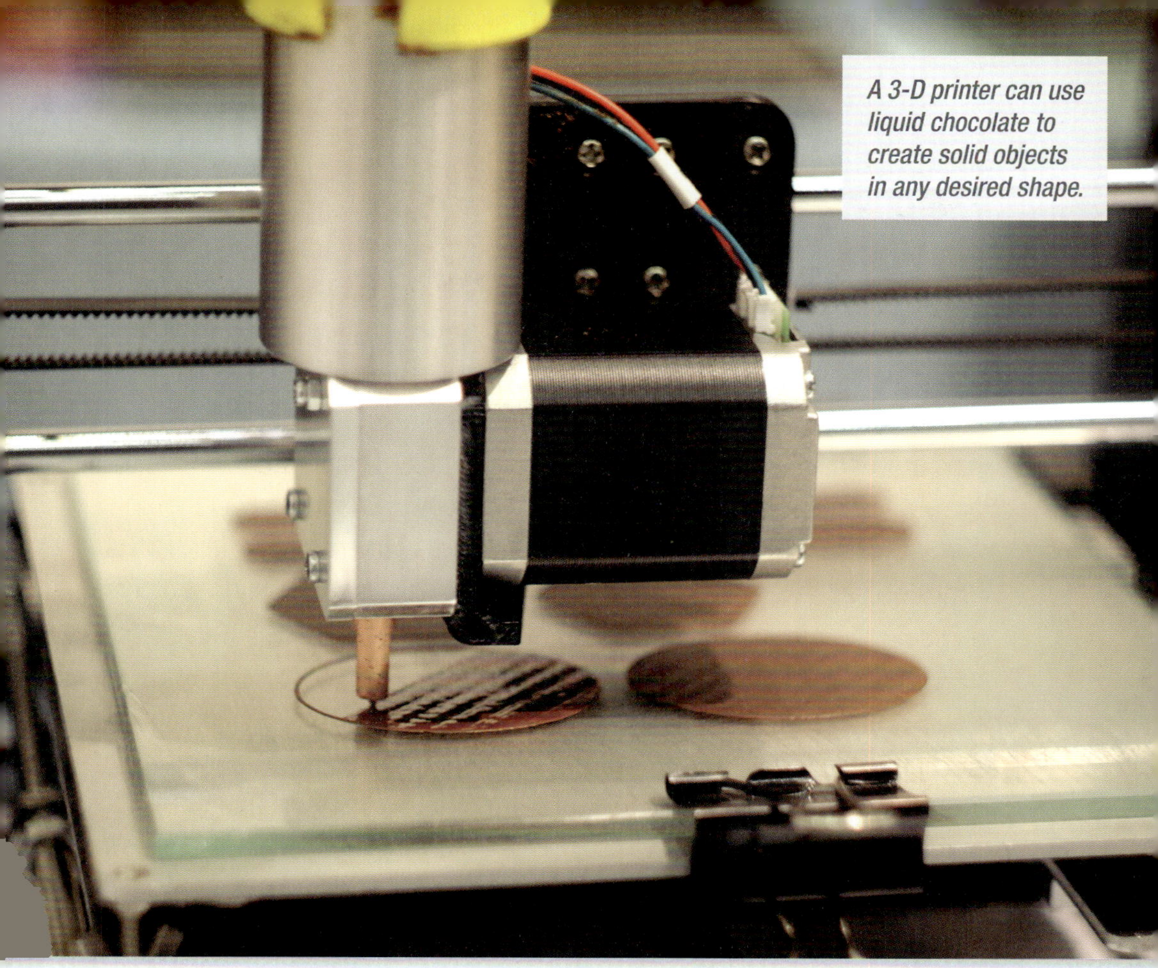
A 3-D printer can use liquid chocolate to create solid objects in any desired shape.

more durable than a traditional refrigerator. It also features USB ports and Wi-Fi connectivity, which allows an owner to operate it via a computer or smartphone.

Not only are appliances being 3-D printed, as incredible as it may seem, food can be produced in a 3-D printer as well. "We all remember the replicator, the Star Trek food manufacturer, able to turn any molecule into edible food and whole dishes," recalls Carlota Valdivieso, an editor of the online newsmagazine 3Dnatives. "This time we are not in a sci-fi movie!" Several companies, she adds, now make 3-D printing machines that can produce "chocolate, pasta, sugar and even more [food types and] dishes. The possibilities are almost unlimited."[23]

The key to 3-D printing food that tastes good and has the consistency of regular food, experts say, is finding the right raw materials in the first place. In theory, of course, any fluid or

Many 3-D Printed Guns

Guns of various types are among the many items that can be found in some American homes. According to an international study published in 2017, there are about 1.1 million registered guns in the United States, while unregistered ones number about 392 million. Of those unregistered guns, an undetermined number were made using 3-D printing technology. Printed guns are illegal in most states, so the vast majority of individuals who make those weapons do so in secret. One of them goes by the name Ivan the Troll. "I'm one of many, many like-minded individuals who're doing this sort of work," he says. He contends that he knows of at least one hundred others who actively work on 3-D printed gun technology and suggests that there are thousands more around the country. Members of this loose network of 3-D gun enthusiasts regularly communicate to compare notes and encourage and support one another. Ivan says that he feels he has a right to own as many guns as he desires. Also, he worries that the government might someday try to confiscate all guns, including the registered ones. In a defiant tone, he states, "Take it from me as someone who's printed a gun. Making a slam-fire shotgun is 100 times easier, 100 times quicker, and about 100 times cheaper than printing a [regular] gun. For eight dollars I can pop down to Home Depot and build a shotgun."

Quoted in Jake Hanrahm, "3D-Printed Guns Are Back, and This Time They Are Unstoppable," May 20, *Wired*, 2019. www.wired.co.uk.

plastic-like material can be fed into such a printer. The most simplistic example is chocolate. Liquid chocolate goes into the printer, and solid chocolate in any shape desired comes out.

Far more complex and tricky, however, is making a slab of steak or beef in a printer. The company Nova Meat, based in Barcelona, Spain, hit on a recipe for the raw material that contains a mixture of rice, peas, seaweed, and various spices. All are measured in precise proportions so that together they have a beefy taste. The Nova Meat's printer takes these ingredients and lays them down in a crisscross design that mimics the pattern of proteins in the muscle cells of cattle. "This strategy allows us to define the resulting texture in terms of chewiness and tensile and compression resistance," says Nova Meat executive Guiseppe Scionti. In this way, he goes on, it is possible to copy "the taste and nutritional properties of a variety of meats and seafood, as well as their appearance."[24]

The Reality of Printed Clothes

In a house that itself was 3-D printed and in which the furniture, appliances, and food the residents eat are printed too, it is not surprising that the clothes they wear could hypothetically be printed as well. The reality, however, is that as late as mid-2020, 3-D printed clothes were mostly restricted to the realms of high fashion and art.

tensile resistance
The resistance to being stretched

The reason that 3-D printed clothes are not yet common in people's closets is that printing clothes is still an extremely expensive process. Even a fairly simple garment requires many hours to be printed and assembled. Each extra hour of production adds to the price, of course, and that puts printed clothes out of the reach of average consumers.

A clear example is the world's first 3-D printed jacket, created by noted fashion designer Danit Peleg. She offers the item on her website for the price of $1,500; also, only one hundred jackets were printed, making them part of a limited production run. For an added sum, Peleg offers to customize each jacket. Nevertheless, Peleg herself predicts that the prices of printed clothes will decrease considerably in the 2020s.

Miami-based fashion designer Julia Daviy agrees. "Everyone wants faster 3D printing" of clothes, she says, "and I'm not an exception. I dream about 3D printers that are able to create high-elastic durable clothes 3D-printed as one single piece without the need for long post-processing and trimming. A much bigger 3D printing volume is essential,"[25] and then the cost of such clothes will come way down.

By the time inexpensive 3-D printed clothes become widely available, it is probable that almost any item in an average home will either be printed or have a printed version available to buy in stores or online. Thanks to the emergence of 3-D printing, people's everyday lives are beginning to undergo a sea change as big or bigger than the introduction of central heating, electric lights, and refrigerated food. According to Joseph Flynt, a reporter and writer for the online magazine 3D Insider:

> Technology is rapidly evolving and we evolve along with it. We become comfortable in our own homes, we buy things with the push of a button, and some of us rarely go outside to face any type of weather due to online shopping. I wonder how long it will take until all of us won't have to leave our houses? Will we enter the Star Trek area and simply materialize products out of thin air? I think that with 3D printing, we just made our first step into that future, and soon we will turn fiction into reality.[26]

Many Marvelous Medical Uses

Thanks to the process of 3-D printing, something that many people hailed as a miracle in 2004 appeared to be doubly miraculous in 2020. The saga of brave Luke Massella and his bold doctor Anthony Atala continues and remains one of the lasting triumphs of modern medicine. Back in 2004, when he was ten, Massella's outlook on life was bleak. Partly as a result of his having the spinal deformity spina bifida, his bladder was malfunctioning, which was causing his kidneys to fail. "I was kind of facing the possibility I might have to do dialysis for the rest of my life," he recalls. "I wouldn't be able to play sports, and have the normal kid life."[27]

dialysis
Blood purification via a machine

Fortunately for Massella, in stepped Atala, then on the staff at Boston Children's Hospital. Taking some cells from the boy's defective bladder, Atala used a modified ink-jet printer to create a new bladder and, in a fourteen-hour-long surgical procedure, implanted it in the patient. If the new organ had lasted only a year or two, all involved would have seen it as a resounding success. Yet in 2020, sixteen years later, Massella's bladder was still functioning perfectly, and he had needed no other surgeries during all those years. "Pretty much I was able to live a normal life after [that single operation],"[28] he remarked in a 2019 interview.

Atala has performed a few similar transplants of 3-D printed organs since his groundbreaking surgery on Massella. This use of 3-D printers remains experimental, and Atala and a number of his colleagues believe that continuing research is necessary. They note that about 20 Americans die each day waiting for organ transplants and that at any given moment some 113,000 people await such lifesaving operations. The hope is that fast-improving printing technology will help reduce the shortage of replacement organs.

Physicians around the world are also enthusiastic about the many other medical uses for which 3-D printing shows promise. In addition to vital organs—such as bladders and pancreases—all sorts of body parts, including ears, bones, and even skin, can in theory be printed. In addition, experiments with printing diverse medicines have been successful, as have attempts to print special surgical tools.

Overcoming Tissue Rejection

The holy grail of 3-D printing in the medical field remains the ability to simply print a new organ when one inside a patient is no longer working. Daniel Freedman of the State University of New York at New Paltz School of Science and Engineering explains the basic science behind the idea:

> There are a lot of people working on printing live cells. The idea is that an organ is basically just a bunch of different cells arranged in a very particular way. If you can get all of those cells in the right place with a matrix that will support their growth, then in theory—I'm no biologist—but you might be able to recreate an organ or recreate something that could function as an organ. People have had some success with 3D printing live cells and getting them to grow afterwards.[29]

The success to which Freedman refers includes isolated cases in which doctors were able to print and insert an organ and have the organ function relatively normally for months or years. The most celebrated example remains Atala's replacement of young Massella's bladder in 2004. One reason that Massella's new bladder remained viable for so long was that his body did not reject it. Such rejection of foreign tissue is one of the chief reasons why making and implanting organs this way fails more often than it succeeds.

matrix
The underlying structure of an object, social institution, or society

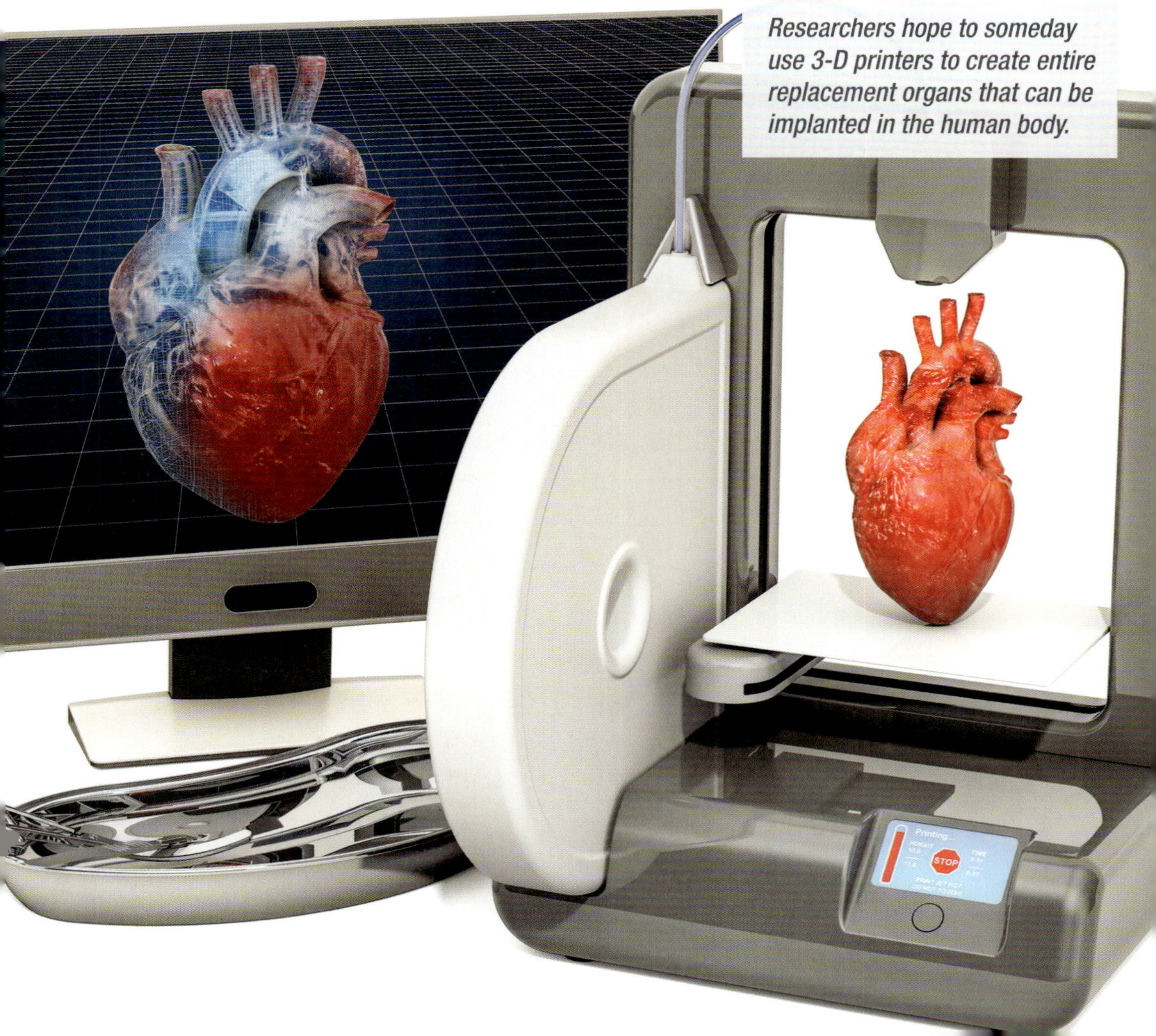

Researchers hope to someday use 3-D printers to create entire replacement organs that can be implanted in the human body.

Atala made sure to include some of Massella's own cells in the biomaterials he fed into the 3-D printer. So when the patient's body detected the new organ, it recognized those cells and did not reject the bladder. Regrettably, in the years since Massella's surgery, such transplants of 3-D printed organs have been hit and miss. Some patients' bodies have rejected the organs even when their own cells were used to make them.

biomaterials
Substances that are part of a living thing, such as skin or blood cells

Mimicking Vascular Networks

Another recurring problem is that to survive, a transplanted organ must regularly receive fresh blood from the circulatory system. To that end, all such transplants rely on the proper functioning of a complex network of tiny blood vessels that exist within and around those organs. Scientists refer to them as vascular networks. If the patient's vascular networks do not grow back quickly after a transplant, the organ will fail even if his or her body does not reject it outright. Massella's new bladder was a success in large part because the vascular networks surrounding it grew back, whereas many other patients have not been as fortunate.

vascular
Relating to the movement of a body fluid, such as blood

A major breakthrough in this area took place in 2018 at Boston's world-famous Brigham and Women's Hospital (BWH). Researchers there were able to 3-D print, or bioprint, a mass of tiny tubular structures that look and act very much like the vessels in a vascular network. In their natural state, such vessels are highly layered, according to Yu Shrike Zhang, one of the chief scientists who worked on the BWH project. That is a major reason why such a network is so complex.

Fortunately, Zhang and his fellow researchers realized, creating a layering effect is one of a 3-D printer's strongest abilities,

Printing New Surgical Tools

One of the more important ways that 3-D printing technology has made it easier for surgeons to operate on patients is through the production of new surgical tools. In their fluid states, metals readily lend themselves to the 3-D printing process, so it is possible to print all manner of metal tools. In addition to making those implements faster and cheaper than with traditional manufacturing methods, 3-D printing lends itself to creating objects with odd shapes. For this reason some surgeons turned to printing technology to make a tool for so-called keyhole heart surgery. This procedure involves making a tiny incision in the heart and passing an instrument through that "keyhole"; the idea is for that tool to hold that section of the heart still long enough for surgeons to finish the operation while the heart is still beating. According to a spokesperson for the 3-D printing company 3DEO, based in Gardena, California:

> In a period of three months, and for a total development cost of under $20,000, a partnership between a British heart surgeon and a 3D printing organization yielded a totally safe and effective new device. Made of stainless steel, the instrument delivered improvement in 90 days that could have taken a decade with traditional methods of prototyping. Stories like this will become much more common in the future, as [3D printing] technologies become more available and advantages of 3D printing for surgical devices become more widely recognized.

Quoted in 3DEO, "Metal 3D-Printed Surgical Instruments: Predictions for the Future," January 31, 2019. https://news.3deo.co.

since by design it builds objects one layer at a time. "The vessels in the body are not uniform," Zhang states, but vary from one layer to the next. To reproduce that effect as closely as possible, the researchers fitted the printer with a custom nozzle that allowed them to spray out the almost microscopic tubular structures up to three layers at a time. "This bioprinting method," Zhang explains, "generates complex tubular structures that mimic those in the human system with higher fidelity than previous techniques."[30]

Generating a replica of a vascular network is one thing, but successfully connecting it to a 3-D printed organ is quite another. So for the moment, doctors around the world are reproducing and trying to build on BWH's achievement. The expectation is that this breakthrough has brought the medical world several steps closer to finding its holy grail. In the words of Katherine Plewa, whose blog reports on trending 3-D printing news, "This revolutionary development proves that we're just a few steps away from fully functional 3D printed human organs."[31]

Replacing Body Parts

Besides ardent attempts to 3-D print internal organs like bladders and pancreases, researchers around the globe are working on the simpler but still difficult challenge of making replacement body

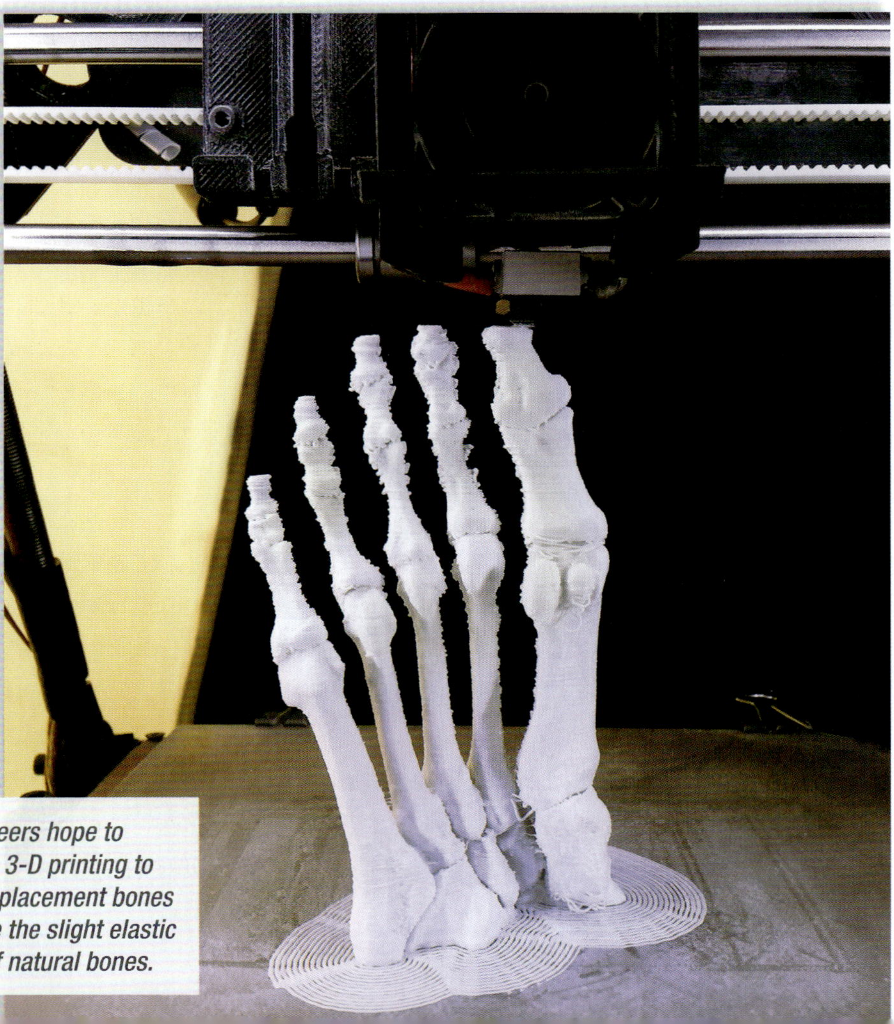

Bioengineers hope to soon use 3-D printing to create replacement bones that have the slight elastic quality of natural bones.

parts. The key factor is that these parts need to be functional. For example, for decades people have been making artificial ears to replace those of people who lost their ears in battle or as the result of disease. Those replacement ears have been composed of a variety of materials, from wood to rubber to plastic. In most cases, however, even the ones that were made on 3-D printers were not functional; rather, they served only to camouflage the loss of the original ear.

In 2018 scientists at Princeton University in New Jersey managed to 3-D print a so-called bionic ear, which not only looks realistic but also functions somewhat like a real ear. The fleshy part of the ear is made of calf cells that hardened into cartilage during the printing process. Hidden within that material is a small coil antenna composed of microscopic silver particles. The antenna, which picks up radio signals, can in theory be attached to a patient's nerve endings in the same manner as a cochlear implant, which converts sound into an electrical signal that human nerves detect. The Princeton team expects this new 3-D printed technology to begin to be used for human patients as early as 2023.

Another category of body part for which 3-D printing shows promise is animal and human bones. Both replacements for and repairs to broken bones are required by the tens of thousands each year in the United States alone. Responding to this need, a team of bioengineers at Northwestern University in Chicago began using a 3-D printer in 2018 to create artificial bones that have an elastic quality, a trait found in natural bone growth and regeneration. The researchers gathered quantities of a human bone mineral, hydroxyapatite, and mixed it with a plastic-like polymer. Although that mineral is brittle by itself, the polymer gives it the desired flexibility. The scientists implanted the printed bones in test animals, with positive results, and human trials will ensue in the near future.

Making artificial skin to aid burn victims is another up-and-coming medical goal involving 3-D printers. From 2018 to 2020 teams from the Canada's University of Toronto and Toronto's Sunnybrook Hospital collaborated on a special printer designed to make sheets of replacement skin for patients who suffer burns covering large parts of their bodies. "With big burns," says cosmetic surgeon Marc Jeschke, who heads the Sunnybrook team, "you don't have sufficient healthy skin available, which could lead to patient deaths." To help fight bodily rejection, the printer uses cells taken from the patient and can make large patches of skin in only a few hours. Jeschke predicts that the device will be ready for use in hospitals by 2025. "Once it's used in an operating room," he says, "I think this printer will be a game changer in saving lives. With a device like this, it could change the entirety of how we practice burn and trauma care."[32]

Making Personalized Medicines

In contrast, it will not be necessary to wait for years to see 3-D printers creating pharmaceuticals, or medicines. The technology to do so is already in place and offers a number of advantages for both doctors and patients. Perhaps the biggest of these is the ability to make personalized medicines; that is, drugs tailored to individual patients who need specialized doses of such substances. "There is a real need for new manufacturing systems to prepare personalized medicines," states Maria-Luz Couce, a physician and researcher at the University of Santiago de Compostela in Galicia, Spain. "The potential applications of 3D printing for rare diseases are countless and it could help to make the medicines safer and more attractive to children,"[33] she says.

Indeed, pediatricians, doctors who specialize in treating children, have long faced the difficulty of prescribing doses of medicine that can vary widely according to the age and weight of their young patients. Clearly, a five-year-old girl who weighs 40 pounds (18 kg) needs a different dose of a given drug than a twelve-year-old boy who weighs 90 pounds (41 kg). The problem has been well known in the pharmaceutical industry, or big pharma, for decades. Namely, making pills containing even one dose of a drug using traditional methods is expensive because it requires setting up a complex production line. Making five, six, or more separate

3-D printing allows pharmaceutical companies to manufacture medicines in doses that are tailored to the needs of individual patients.

doses of that medicine—for multiple children's age groups—can therefore be extremely costly for a manufacturer. As Mark Jones, editor of the technology news website TechHQ, observes, "Every medicine is created to a certain standard, and making pills smaller means reducing the dosage. If creating one medicine can cost billions in big pharma, making two variants would be an extremely difficult and costly process."[34]

By contrast, 3-D printing offers a way around this challenge. A printer can fairly quickly and inexpensively be adjusted to make any number of different doses of a medicine. According to Jones:

> Using a 3D printer allows pharmacists and manufacturers to rapidly design and create pills to different specifications, small or large, longer absorption or slower to meet the patient's needs. Once printed, robotic systems can fill up capsules to the right dosage. . . . With 3D printing, medicines can be customized, allowing children, adults, and the elderly to be treated individually with the right dosage.[35]

Printed Prosthetics for Pups

3-D printing has shown promise for making prosthetic limbs, not only for human patients but also for canine ones. In 2017, for instance, a four-year-old greyhound named Millie became the first dog in Australia to be fitted with a 3-D printed prosthetic leg. When she was only eleven weeks old, she was rejected by a local dog-racing organization due to problems with a malformed and very painful front paw. A compassionate couple—the Dieppes, who had in the past rescued other pups—took Millie in. Not long afterward, the greyhound had to have the ailing leg amputated, and the Dieppes ordered a prosthetic leg. It did not fit properly, and neither did two other prosthetic devices the couple bought at considerable cost. At that point a tip from a friend led the Dieppes to the British 3-D printing software company Autodesk, which for only a few hundred dollars printed a custom plastic leg for Millie. She now jumps and runs quite comfortably. "It really just allows her to be free," Mr. Dieppe says. "And for us as her parents," it is nice "not to worry about her damaging her leg. She likes the fit and feel of this one much more than any of her other prosthetics."

Quoted in Hanna Watkin, "Millie the Greyhound Is First Dog in Australia to Receive a 3D Printed Prosthetic Leg," All3DP, October 18, 2017. https://all3dp.com.

Simulated Surgery and Training Doctors

Still another important medical use of 3-D printers, one that is becoming increasingly routine in medical schools and hospitals worldwide, is to help train doctors, especially surgeons. The sad fact is that errors made by doctors and other medical personnel are sometimes deadly. In 2019 a study by David H. Gorski, a surgeon at the Karmanos Cancer Institute in Detroit, Michigan, concluded that such errors directly cause roughly 5,200 deaths annually in the United States and indirectly contribute to another 108,000 fatalities. "One death from medical error is too many," he states. "We can do better. We *should* do better."[36]

The chief way that 3-D printing promises to reduce these unfortunate numbers is by creating highly detailed, realistic-looking three-dimensional models of various parts of the human body.

Surgeons, both trainees and veterans, use these models to practice on. Before the advent of this technology, doctors routinely practiced on either body parts from human cadavers or parts from dead animals—often pigs, some of whose internal organs closely resemble human ones. These methods entail various practical problems and in some cases raise ethical questions.

In comparison, the 3-D printed models not only eliminate such problems but also provide doctors with a more realistic practice experience. In many cases the models can even be customized to specific patients. A surgeon can take ultrasound or other images of a patient and upload them to a company that prints the models. In only a few days the doctor receives the completed models and can begin his or her practice session. Those simulated body parts are becoming more and more sophisticated; some of the most detailed ones can even mimic the kind of bleeding that the surgeon can expect during the upcoming procedure.

Considering the success to date of various applications of 3-D printing to the medical field—a list that is constantly growing—the future of printing technology in health care promises to be a bright one. In the words of mechanical engineer Farai Mashambanhaka, a leading expert on that technology, "3D printing in medicine is accelerating the health care industry at an impressive rate. The result is more comfort for the patient, better understanding for the doctor, and an easier time for the pocketbook. The precision, speed, and low cost is paving the way for a more intelligent and joyful approach to healthcare."[37]

New Life for Ancient Artifacts

In the popular films about his exploits, the fictional archaeologist and adventurer Indiana Jones used atypical tools—a bullwhip and a pistol—to find and preserve priceless artifacts, such as the lost ark of the covenant. Today real archaeologists and the museums where their finds end up are using a no less unusual and unprecedented tool—the 3-D printer. It not only preserves ancient artifacts that archaeologists have unearthed but can also piece together objects that ceased to exist thousands of years ago.

This astounding ability was first demonstrated in 2012 at the Harvard Semitic Museum, which displays precious artifacts from the ancient Middle East. Archaeologists and curators Joseph Greene and Adam Aja approached a task so daunting that all the museum curators before them had simply given it up as a lost cause. Tucked away in its basement vaults, the museum had a jar containing several hundred pottery fragments. Excavators had found them decades before in the ruins of the ancient Mesopotamian city of Nuzi in present-day Iraq. Most experts had guessed that these pieces once formed a ceramic lion. But it was difficult to tell for sure, because invading Assyrians had smashed the

curator
A professional who maintains a museum and safeguards and displays its contents

sculpture to bits when they captured Nuzi some three thousand years ago.

Greene and Aja were aware that a relatively new tool—the 3-D printer—could reproduce models of any object. The question was whether that device could make a copy of an object using only its disjointed fragments as reference points. Carefully and painstakingly, the two scientists photographed the fragments from hundreds of different angles. They then loaded those images into a computer, which pieced them together into a two-dimensional image of what was indeed a lion, just as had been suspected. The final step was to feed that digital image into the 3-D printer. Using a special foam that hardens into a ceramic-like material similar to the kind making up the fragments, the device generated a perfect model of what the sculpted lion originally looked like.

photomodeling
A process in which scientists and/or museum curators use a 3-D printer to piece together the broken fragments of a display object

This new process, which became known as photomodeling, created a stir in the worlds of archaeology, history museums, and the arts. The Harvard team had provided an important addition to the toolbox that members of those fields use to connect with and preserve humanity's past. Greene himself said in an interview, "3-D imaging [with] or without printing is a perfect way to study, conserve, share, and teach using [ancient] objects."[38]

Restoring Historical Artifacts

Considering those qualities of 3-D printing, it is not surprising that other museums in the United States, Europe, Asia, and elsewhere were quick to incorporate it. According to Charlotte Coates, who reports on exhibits and activities at museums worldwide, "There are many museums around the world already embracing 3D printing. Curators have been exploring how it can add value to their collections."[39]

One of the more common ways those curators have used 3-D printing so far is in repairing and restoring damaged historical artifacts. Every museum has at least some ancient statues, paintings, tools, weapons, and other artifacts that are missing parts. Sadly, such damaged pieces are often the ones that curators decide to remove from their exhibits and consign to the museum's storage area, where the public can no longer view them.

With a 3-D printer, however, museum personnel can restore the missing parts, typically using techniques similar to those Greene and Aja used at Harvard to restore the ancient Mesopotamian lion. Spanish archaeologist Néstor Marqués is one of the world's leading experts in this kind of restoration. He employs 3-D printers to repair ancient Roman pottery and marble artifacts for museums around the globe. In 2018 alone he restored more than fifty display pieces and worked on nearly the same number in 2019.

In some cases, Marqués supplied missing arms, hands, heads, noses, and other body parts. On occasion he has restored up to 80 percent or more of an ancient vase, bust, or statue. He begins by making an exact copy of the damaged artifact, prints

Museum curators are using 3-D printers to create restorations of ancient artifacts that have sustained damage over the centuries.

Consuming Art in the Literal Sense

The British Museum in London is one of the premier museums in the world. With more than 8 million items in all, it also has one of the world's largest collections of historical artifacts. Although people travel from all around the globe to visit the museum, its officials have long searched for ways to share some of its exhibits with people who are unable to make it to London. The advent of 3-D printing presented those curators with a unique way of reaching out to the human community. The museum launched an ambitious 3-D printing program in 2017 that places images of selected exhibits on a special website. By clicking on an image, anyone with access to a 3-D printer can make a perfect replica of that object. The first piece the museum put online was one of its signature artifacts—the famous Rosetta stone. Discovered in Egypt in 1799, this stone slab shows a decree released by an Egyptian king in 196 BCE. The message is shown in three languages—Greek and two forms of ancient Egyptian. Because early modern archaeologists could easily read Greek, the stone allowed them to decipher ancient Egyptian hieroglyphics. 3-D copies of that artifact, along with replicas of the other artifacts released to the website, are also available in the museum's gift shop. In a lighthearted twist, several, including small copies of the Rosetta stone, are available in chocolate form, which allows visitors to consume—literally—great art.

the needed replacement parts, and then glues those parts to the copy. Finally, he carefully uses plaster, paint, and other materials to make the copy look exactly as the original once appeared. Marqués and other experts do not alter the damaged original. Most often the museum displays both the damaged original and the restored copy beside each other, which clearly shows observers how much time and random circumstance can change ancient objects.

One of the added benefits of using 3-D printers for this sort of restoration work is that multiple museums can benefit. For example, one of Marqués's 2018 projects was a twenty-three-hundred-year-old statue that he pieced together from twenty-one pieces. When he finished gluing and painting the printed replica for the National Archaeological Museum of Spain in Madrid, he printed a copy of it and sent it to a museum in the Spanish town of Vilamuseu. That way both facilities are able to display copies of the object.

Making the Past Come to Life

3-D printing technology also allows replicas of priceless artifacts to leave museums in other ways. One of the more popular ones involves making the copies available to people who for one reason or another are unable to visit a museum in person. In 2018, for example, the famed Victoria and Albert Museum in London teamed up for an innovative outreach project with London's Great Ormond Street Hospital for Children. The hospital's staff has long used art projects as a tool to enrich patients' lives and thereby help them either prepare for surgery or recuperate faster afterward.

To that end, the Victoria and Albert staff hired an art restoration company to make 3-D printed copies of several sculptures and brought them to the hospital. There, more than a dozen children were housed in an isolation ward as they awaited bone marrow transplants. About the project, Victoria and Albert curator Alex Flowers remarked, "Taking the museum out of the walls of South Kensington [London] to those who can't make a visit is so important for making sure that culture and our collections are accessible to everyone. Visits to hospitals can be scary times for both patients and their families and by bringing in creative activity we want to create a meaningful distraction and engagement."[40]

tactile
Related to the sense of touch

Not only were the young patients able to see the sculptures, they could also touch them. That added a tactile dimension to art and history that in the past was not allowed in museum settings, since visitors were routinely greeted with signs warning them not

Some museums are using 3-D printers to make copies of delicate artifacts so that blind people can enjoy exhibits through their sense of touch.

to touch the exhibits. Today, both inside and outside of museums, 3-D printing technology increasingly allows people to touch articles from humanity's past, in very real ways making both those artifacts and the past itself come to life.

Thanks to 3-D printers, New York's Metropolitan Museum of Art (the Met) has also instituted touchable displays in some of its exhibit halls. By mid-2019 the Met already had more than seventy 3-D models on display, including sculptures, busts, standing statues, and furniture items. Visitors are encouraged to touch these replicas to their hearts' content. One of the museum's printed guides states, "The MediaLab at the Metropolitan Museum of Art has been working with creative technologists and 3D enthusiasts, using . . . hand-held 3D scanners, to render objects in our collection into 3D models that we can share with the world. We encourage everyone to use our content, which represents the world's cultural heritage, to create their own creative works."[41]

Touchable exhibits and artifacts have proved especially helpful to various disabled people, particularly blind individuals. As Marqués, whose printed copies now grace museums worldwide, puts it,

Printing the Remains of Ancient Disaster Victims

University of West Florida archaeologist Kristina Killgrove was part of a group of American and Italian excavators who visited the Bay of Naples in western Italy in 2017 and 2018. There in 79 CE the volcano Mount Vesuvius erupted with enormous violence. The disaster buried several towns and villages, the best known being Pompeii and Herculaneum. A number of wealthy estates and villas in the area were also erased from view, including one called Oplontis, which the team investigated in considerable detail. The scientists found the skeletons of some fifty people in a single building; the assumption is that on that terrible day they were likely awaiting a rescue by the Roman navy that never came. Killgrove and her colleagues were armed with advanced archaeological tools, including 3-D printers. With these devices they made three-dimensional copies of many of the human remains. By printing "individual skulls and the skeletons of more than a dozen people who lay where they died more than two millennia ago," Killgrove explains, "[we] hope to clarify the processes by which the people were killed and reconstruct what their lives, and faces, looked like." The 3-D printed skeletons, she adds, are "incredibly useful for understanding the ancient Roman world . . . because it represents a cross-section of people cut down in their prime, unlike a cemetery that holds the remains of people who most often died of disease."

Kristina Killgrove, "Archaeologists Reveal 3D Skulls of Ancient Romans Killed by Eruption of Vesuvius," *Forbes*, January 22, 2018. www.forbes.com.

"Accessibility for disabled people is a great advantage of 3D printed replicas. Imagine a blind person in a museum. Everything is behind the glass, and they cannot touch anything, so they can't enjoy [the exhibits]. 3D printing ensures they get a great heritage experience by knowing they are touching an exact replica of the object."[42]

Dual-Purpose 3-D Archives

Still another practical use that a number of museums have found for 3-D printers is creating archives, or backup reproductions, of their entire collections. Such printers are incredibly sensitive and capture more detail than the human eye can. This makes them the perfect vehicle for copying the items in a museum collection and thereby preserving them for posterity. All major museums have made two-dimensional photographic archives in the past, but traditional photos cannot reproduce the extra dimension of depth that makes 3-D printing so special and desirable.

The Met, the British Museum in London, and the Smithsonian Institution in Washington, DC, are only three of the world-famous museums that are currently in the process of archiving their collections in this manner. All see these archives as being dual purpose. On the one hand, they catalogue those collections for the ages; on the other, they provide a way for museums to share with the public their complete collections, including the many items that are normally out of sight in storage.

The Smithsonian project is particularly daunting because of the sheer size of that facility's collection. It is estimated to contain more than 155 million separate items. Fewer than 5 percent of them have been 3-D printed so far, but the museum is wasting no time in sharing what has been catalogued. In early 2020 it released around 2.8 million high-resolution 3-D images on an online site that anyone can access for free. Museum officials said they planned to release another 200,000 images later in 2020 and that the project would continue to its completion sometime in the 2030s. One of the project's overseers, Effie Kapsalis, said in February 2020, "Being a relevant source for people who are learning around the world is key to our mission. We can't imagine what people are going to do with the [online] collections. We're prepared to be surprised."[43]

Preserving Historical Sites

In addition to reproducing items inside museums, 3-D printing technology has already shown much promise for mapping and preserving the layout of the archaeological sites where many of those items were originally found. Unfortunately for researchers, and human society in general, such historical locations do not last forever. They constantly deteriorate under the onslaught of modern urban expansion and human-made pollution, as well as earthquakes, floods, and other natural disasters.

Now these original ancient, medieval, and early modern excavation sites can be 3-D scanned, and the images will survive for further study even when the sites themselves are altered or destroyed. Several dozen historical sites have already been preserved in this way. One of those projects, which began in 2016 and remains ongoing, involves making extensive 3-D scans of an area just off the coast of northwest Scotland. There in the early 1690s a Dutch trading vessel, the *Crowned Raven*, sank as it was about to dock. British archaeologists worried that the remnants of the ship and their unique submerged layout near the coast might be largely lost before enough money could be found to raise them to the surface.

The 3-D images created to date have accurately mapped the site and stunningly captured the original colors of the cannons, storage casks, and other remains of the wreck. The project's chief excavator, John McCarthy, told an interviewer, "It's been a fascinating process to transform the light captured in the photographs and the sound captured by the sonar sensors back into solid objects through the 3D printing process. We are very excited about the potential for this technology to help us to show the wider community what it's like to visit the site without having to learn to dive or even get your feet wet."[44]

Printing Dinosaur Bones and Sharing Data

Long before there were ships to sink or even people to build those vessels, all manner of prehistoric beasts walked the earth, among them dinosaurs. Using traditional methods, preserving, shipping, and reassembling those creatures' bones for museums can be costly. For that reason, in recent years more and more paleontologists are turning to 3-D printing technology.

paleontologist
A scientist who digs up and studies the remains of ancient creatures

Paleontologists have begun using 3-D printing technology to preserve dinosaur bones by creating exact replicas that can then be reassembled into a complete skeleton.

A fascinating example is the work being done by Art Chadwick and his team at Southwest Adventist University (SWAU) in Keene, Texas. Although small compared to several other US universities that teach paleontology, SWAU has an amazingly large collection of dinosaur bones—amounting to more than twenty thousand in 2020 and still expanding.

Much of that collection came from a project that began in 2000 and more than two decades later is still going. Initially, a colleague asked Chadwick to take a look at some dinosaur bones on a ranch in Wyoming. After Chadwick arrived there, he recalls, "the ranch owner took me out onto his property, and he drove his pickup up onto a butte, stopped, and told me to get out." When he climbed out of the truck, Chadwick was startled to see how careful he had to be in taking every step. "It was covered with dinosaur bones!"[45] he says.

In fact, as the years rolled by the SWAU team collected so many bones that it became increasingly difficult to find university money to ship and preserve them all. "We began to realize," Chadwick remembers, that "we needed to be able to [3-D] print bones so that we could reconstruct some of the animals that we were finding, especially as we began to find whole animals."[46] One of those complete skeletons was that of an 8-foot-long (2.4 m) plant-eating creature called *Thescelosaurus*. The team found that to reassemble the creature for display would require breaking several of the precious bones. Yet by using 3-D printing, they could make replicas of the bones and assemble those, while keeping the original bones intact. "We kept [the printer] busy, day and night, for a long time, printing out all those bones,"[47] Chadwick says.

As more bones, including several full skeletons, continue to arrive at the university, the SWAU team members marvel at the enormous amount of data about dinosaurs that the printing process has helped them collect. "To me," Chadwick remarks,

> 3D printing is opening a whole new avenue of sharing information, which is what science is all about. If you're not sharing information, you're not doing science. It's for the general good and advancement of knowledge to share information with your fellow researchers. If you find something [valuable], the best thing in the world you can do is to share it, so that other people can access it. Not just for paleontology but for biology [and knowledge] in general, 3D printers are a boon.[48]

In a few short years, therefore, 3-D printing has revolutionized attempts to preserve nature's and humanity's past. Whether to preserve the bones of ancient creatures, keep intact ancient archaeological sites, or repair damaged museum exhibits, 3-D printers are working overtime to keep that priceless heritage from disappearing forever.

CHAPTER FIVE

3-D Printing in the Final Frontier

"I'll never forget the first time I saw a rocket materialize before my eyes," says aeronautical engineer Bryce Salmi with a touch of awe. "In October 2018, I stood in a small room and watched a massive robotic arm move elegantly around a large metal shape, which was rapidly growing larger as I gazed at it. The arm precisely deposited a stream of liquid aluminum to build up the structure, layer by layer, while two other arms waited, with finishing tools at the ready."[49]

At that milestone, Salmi stood in the Los Angeles facility of the recently formed rocket company Relativity Space. Its mission is to create launch vehicles, or the rockets that initially carry satellites and astronauts into Earth orbit and sometimes beyond into the vast reaches of outer space.

The company calls the launch vehicle whose creation Salmi witnessed the Terran 1. What makes it different from the space rockets of the past is that through the process of 3-D printing it can be made far faster and much less expensively. "It's normally a monumental task to get a single engine designed, fabricated, and on the test stand," he explains. Normally "this process takes ten to twelve months using traditional methods. A major redesign takes almost as long."[50] It also costs many millions of dollars to create a launch vehicle big enough to carry space capsules containing humans into orbit. In comparison, Relativity Space's huge, sophisticated 3-D printer can fabricate a large-scale rocket in only a few days and for the shockingly small cost of less than $100,000.

Not surprisingly, this astounding advance in space hardware has not gone unnoticed in the global space exploration industry. In 2019 four major companies that annually launch multiple satellites into orbit, including Spaceflight and Telestat, signed deals with Relativity Space to use the Terran 1 beginning in 2021. In addition, also in 2019 the US Air Force agreed to allow Relativity Space to erect a launch site at Cape Canaveral, the National Aeronautics and Space Administration's (NASA) renowned Florida space facility. At the same time, NASA provided Relativity Space with a building in which it will manufacture up to twenty-four rockets a year. "Such mass production will represent a revolution in rocketry,"[51] Salmi says.

Lighter, Cheaper Space Hardware

Manufacturing launch vehicles to carry payloads into orbit is only one of dozens of ways that 3-D printing technology promises to revolutionize space exploration. One of the more important is supplying a multitude of parts for rocket engines, space capsules, and other space hardware. Relativity Space's massive printer specializes in fabricating large-scale objects, primarily entire rockets. But space capsules, satellites, space stations, and interplanetary probes are composed of thousands of much smaller mechanical parts. No supplier yet exists where engineers can buy such items in bulk and at reasonable cost. Rather, almost all of these parts are one of a kind. So they must be created from scratch, which is a very expensive and time-consuming process when using traditional means.

payload
Items carried by a spacecraft or aircraft that make up the purpose of the flight

3-D printing offers an unprecedented and exciting alternative to those traditional manufacturing methods. Only one of many recent examples is the Orion spacecraft. According to an October 2019 NASA press release, it is a new NASA spacecraft for astronauts. The spacecraft is an important part of NASA's Artemis program that includes sending the first woman and the next man to the moon by 2024. Orion will carry astronauts into deep space, and then return them home safely to Earth. This spacecraft will take a crew to the Gateway in lunar orbit. The Gateway is a new spaceship that will serve as a place for astronauts to prepare before missions to the surface of the Moon and, ultimately, Mars.[52]

One way that both Orion and the Gateway will differ significantly from prior moon missions is that they will be equipped with new, especially lightweight engines. In the past, the hundreds of parts making up such motors were composed of metal. To be strong enough to withstand the pressures created by the thrust of a firing engine, the metal had to be dense and therefore heavy. Getting all that extra weight off Earth's surface and into space was both difficult and expensive, which in turn limited the size and quality of spacecraft of all kinds.

NASA astronauts Nick Hague and Anne McClain work on electronic gear in the International Space Station's Japanese Kibo laboratory module. 3-D printers are being used to manufacture thousands of the small mechanical parts used in the space station.

In contrast, Orion's engine features more than one hundred 3-D printed parts. Each is made of a plastic resin that is lightweight but extremely strong and durable, as well as far cheaper and easier to produce than any metal equivalents. Moreover, says Scott Sevcik, an executive at the space engineering company Stratasys, the plastic material is less susceptible to damage from static electricity. "In space," he points out, a rocket's or capsule's "materials will build up a charge,"[53] and that excess electricity is potentially dangerous because [discharging] it can start a fire or explosion. "A little spark," Sevcik says, "can cause millions of dollars' worth of damage."[54]

static electricity
An imbalance of electric charges on or within a material

3-D Printing on a Space Station

Aside from many of the parts for the capsules and rockets that travel in outer space, 3-D printing already supplies large numbers of tools and other materials needed to sustain astronauts who live

The engine of NASA's spacecraft Orion features more than one hundred parts produced on a 3-D printer.

in space stations. The personnel who dwell on the International Space Station (ISS), for example, require all sorts of tools, technical gear, personal items, and other objects on a daily basis. To meet this demand, each year NASA, Russia's Roscosmos space agency, and other space forces send more than 7,000 pounds (3,175 kg) of such items to the station. On top of that, over time the ISS astronauts have collected and stored at least 29,000 pounds (13,154 kg) of spare parts and backup gear on board.

In the craft's first few years of existence, every single item either used or stored on the ISS had to be made on Earth and flown into orbit. This fact alone is a potent reminder of why space exploration has traditionally been so enormously expensive. Furthermore, most of the habitats housing astronauts have so far been located in orbit, only around 250 miles (402 km) above Earth's surface. And shipping thousands of pounds of vital materials and spare parts from Earth to astronauts on long-range missions to the moon, Mars, and beyond will be logistically unrealistic and far too expensive.

The advent of 3-D printing has greatly changed that situation, space experts say. As an experiment, in 2014 NASA teamed up with a California-based company, Made in Space, to create a 3-D printer for the ISS. The crew of the station received the device late that year and at first had to determine whether 3-D printing would even work in near-zero-gravity conditions. It soon became clear that the process does work in a weightless environment as well as on Earth's surface.

In the years that followed, astronauts continued to use the printer to see whether such devices could produce enough necessary tools, spare parts, and other items to significantly reduce costs over time. The first useful printed object the station's astronauts made was a wrench. A team on Earth, supervised by 3-D printing expert Niki Werkheiser, created the computer specifications for the tool and sent them via NASA's electronic link to the astronauts on the ISS. The astronauts then fed the information into the printer, along with some raw plastic, and printed out the wrench in about four hours. "We wanted to work this just like we would for tools that the astronauts will 3-D print and use on the station," Werkheiser later recalled.

> This wrench will not be used in space, but what if it were a tool the crew needed? We are breaking new ground not only in the way we manufacture in space but also in the way we operate and approve space hardware that is built in space, rather than launched from Earth. If you can transmit a file to the station as quickly as you can send an email, it opens up endless possibilities for all the types of things that you can make [in space].[55]

These possibilities were aptly demonstrated in the years that followed. By early 2020 the ISS's crew had made hundreds of objects with the printer, including tools and other objects that would normally have cost thousands of dollars to ship there on rockets. Moreover, in late 2019 Made in Space updated the system by sending the station a specially designed recycler. Thereafter, the astronauts were able to recycle all unused plastic objects and feed them into the printer to make new ones.

Space Filled with Building Materials

According to experts at NASA and other space agencies worldwide, making objects that are usable for astronauts living in space represents only the beginning of the marriage between 3-D printing and space exploration. Already, right here on Earth, such devices are being used to build entire houses, and that is something that will happen in space as well. Planetary bases and even entire colonies will, over time, be erected on the moon, Mars, and likely elsewhere in the solar system. Workrooms, science labs, living quarters, and storage facilities will all be needed.

The main reasons that 3-D printing will be the technology of choice to make those habitats are both logistical and monetary. Bringing all those building materials up from Earth's gravity will be

3-D Printing in Microgravity?

At first, space scientists were not sure whether it would be possible to use the 3-D printing process in space. The worry was that perhaps microgravity, or the near weightlessness of people and objects in Earth orbit, might somehow impede that process. For that reason, the 3-D printer developed by the company Made in Space in 2014 and 2015 for use on the ISS was thoroughly tested by astronauts on the station. After printing a number of objects, they found that microgravity had no significant ill effects on the 3-D printing process. In the words of NASA engineer Tracie Prater, "Our current hypothesis based on work so far is that this particular 3D printing process is not affected in a significant way by microgravity. . . . As with any process, it is important to have good process control. The idea with the technology demonstration was to show that you can do this in microgravity, and the only way to test that was operating over time on the station."

Quoted in Melissa Gaskill, "Solving the Challenges of Long Duration Space Flight with 3D Printing," NASA, December 16, 2019. www.nasa.gov.

much too difficult and costly. It makes far more sense, the experts say, to use the materials already available in space. According to Chris Lewicki, a member of ASM International, a group of scientists and engineers who do consulting work, "Instead of manufacturing something in an Earth factory and putting it on a rocket and shipping it to space, what if we put a 3-D printer into space and everything we printed with it we got from space? There are billions and billions of tons of [such building materials] in space."[56]

Indeed, scientists have determined with certainty that the solar system is literally filled with those kinds of building materials. Among them are metals—iron, nickel, titanium, aluminum, and others. These will be mined from asteroids, which exist by the millions between Mars and Jupiter and in lesser numbers in other sectors of nearby space. This is not a fanciful prediction of something that might happen many decades from now. Rather, NASA has already landed unmanned probes on asteroids for future mining missions that are currently in preparation. "By 2025," says an ASM International spokesperson, mining asteroids is likely to "be as commonplace as digging a hole in your backyard."[57]

Other building materials for space habitats will consist of a wide array of minerals. Analysis of moon rocks brought to Earth in the 1970s show high concentrations of silicon, calcium, and feldspar. Large deposits of these and other minerals have also been detected on Mars, and some are sure to exist in the outer crusts of most planetary moons in the solar system.

Designs for space habitats made of 3-D printed metals and minerals have already been formulated. One of the more practical and promising is a group of Mars structures designed by the noted British architectural firm Foster and Partners. As envisioned in the proposal, NASA and other space agencies would first land

Scientists and engineers hope that asteroids like this one will someday be a source of the raw materials, such as iron, nickel, titanium, and aluminum, that will be used by 3-D printers located in space.

robots on Mars that would be directed to do most of the printing and construction. Each shelter, intended to house four astronauts, would be made of Mars regolith—the metal- and mineral-rich local soil. The robots will initially dig a 5-foot-deep (1.5 m) hole. They will then separate the metals and minerals from the extracted regolith and feed the usable materials into a 3-D printer, which will fashion the panels for the walls and roofs. When the astronauts eventually arrive, they will put the finishing touches on the structures.

regolith
The metal- and mineral-rich soil of the moon, Mars, or other space body

3-D Printing of Space Habitats Is Inevitable

Although many shelters and entire communities on planets and their moons will surely be constructed from metals and minerals fed into 3-D printers, another potential building material exists in nearby space in fair abundance—ice. Ice has already been found hiding in the shadows on the floors of some lunar craters. Moreover, huge amounts of water exist on Jupiter's moon Europa, Saturn's moons Enceladus and Phoebe, and other natural planetary satellites.

Fed by water, a 3-D printer can create panels of ice in any shape or thickness that builders require. This was the conclusion that a team of space enthusiasts reached during a recent competition NASA held. The contest challenged 165 teams of participants to develop architectural ideas for employing 3-D printers to create habitats on Mars, and by extension other planets and moons. The stipulation was that all materials that would be printed to build the habitats had to be in situ, or already on Mars.

in situ
Already in existence

The winning team—calling itself Search and Clouds AO—received an award of $25,000 for its concept of the so-called Mars Ice House. Comprising designers Christina Ciardullo, Kelsey Lents, Jeffrey Montes, and other New Yorkers with ties to that

Printing Human Organs in Space

Each day around twenty people in the United States die waiting for an organ transplant. The recently formed biotechnology company Techshot hopes to save most or perhaps all those lives by 3-D printing replacement organs in space. In 2019 Techshot partnered with NASA to test the feasibility of that goal. Although some human organs have been 3-D printed on Earth, in the cases of hearts and other large, complex organs, during the printing process the very weight of the tissues within the organs often makes them collapse and become useless. Techshot executives believe that part of the problem is the constant pull of Earth's gravity on the fragile tissues. Printing the organs in the zero gravity environment on an orbiting space station like the ISS, they suggest, could be the solution to the problem. Techshot's 3-D printer made it to the ISS in May 2019. Rich Boling, the company's vice president, explains that Techshot's foremost goal is to save the lives of as many people needing transplants as possible. The initial testing phase, which could last till late 2020, will determine whether the space-made printed organs are stronger than their earth-made counterparts. "After our test protocols have been completed," Boling told a BBC interviewer in mid-2019, "we'll open the program up to outside researchers who want to use our device." If the tests prove to be completely successful, he adds, cardiac, or heart, patients will become the initial focus of the orbiting bioprinting program.

Quoted in Hanna Watkin, "NASA and Techshot Trying to 3D Print Organs in Space," All3DP, January 24, 2019. https://all3dp.com.

state's universities, the team did not merely propose a theoretical idea. With the aid of several scientists, they actually used a 3-D printer to make sample models of ice structures. A press release from the team states that with the input of consulting scientists, "we have achieved positive experimentation with one to one ice printing. . . . Through an understanding of the physics of [creating ice from water] and the temperature and pressure conditions of the Martian environment . . . we've designed a process to turn subsurface ice into water vapor [and] liquid water in an environment cold enough to print a form in solid ice."[58]

Other recent visions of space habitats have proposed 3-D printing the construction materials from all sorts of substances, including plant matter grown in special gardens on Mars and other planets. Whatever the materials chosen, 3-D printing increasingly lies at the heart of these visions. The ability to make shelters and eventually whole colonies in space, experts point out, will inevitably require manufacturing them via printing technology. One of those experts, Tracie Prater, an engineer at NASA's Marshall Space Flight Center in Huntsville, Alabama, says:

> For long-duration missions, we can't launch everything from Earth. We have to be able to use the materials we have around us. . . . I tend to think of human spaceflight as a long camping trip. If I'm [going to Mars], I can't pack everything I would ever need. [3-D printing in space] enables you to create your own solutions. After all, where would the Star Trek Enterprise be if Scotty [its chief engineer] couldn't make anything?[59]

SOURCE NOTES

Introduction: The Possibilities Are Endless

1. Technology.org, "Scientists Printed a 0.1 mm Tall Michelangelo's *David*," January 20, 2020. www.technology.org.
2. Chris Woodford, "3D Printers," Explain That Stuff!, January 16, 2020. www.explainthatstuff.com.
3. Woodford, "3D Printers."
4. Woodford, "3D Printers."

Chapter One: A Revolution in Transportation

5. Quoted in Carrie Wyman, "Exclusive Interview with the Big Wheel in 3D Printed Automotive Manufacturing," Stratasys, March 5, 2014. http://blog.stratasys.com.
6. Quoted in Wyman, "Exclusive Interview with the Big Wheel in 3D Printed Automotive Manufacturing."
7. Omar Jumaah, "A Study on 3D Printing and Its Effects on the Future of Transportation," doctoral dissertation, Rutgers University, 2018. https://cait.rutgers.edu.
8. Quoted in Megatrends Editor, "How 3D Technology Is Transforming the Auto Industry," *Megatrends by HP* (blog), January 16, 2019. https://hpmegatrends.com.
9. Quoted in Bridget O'Neal, "Three Wheels & Two Seats: Will the 3D Printed Urbee 2 Be Your Car of the Future?," 3DPrint.com, March 14, 2016. https://3dprint.com.
10. Quoted in Tamara Warren, "This Autonomous, 3D-Printed Bus Starts Giving Rides in Washington, DC Today," The Verge, June 16, 2016. www.theverge.com.
11. Quoted in Tekla S. Perry, "First the E-Bike, Next the Flying Car," *IEEE Spectrum*, July 12, 2019. https://spectrum.ieee.org.
12. Quoted in Perry, "First the E-Bike, Next the Flying Car."
13. Quoted in Parmy Olson, "Airbus Explores Building Planes with Giant 3D Printers," *Forbes*, July 11, 2012. www.forbes.com.
14. Marc Prosser, "This Boat Was 3D Printed—and Bigger, Wilder Projects Will Soon Follow," SingularityHub, November 14, 2019. https://singularityhub.com.

Chapter Two: An Entire House and Its Contents

15. Quoted in Michelle S., "Family Moved into Social Housing Project with 3D Printed House," 3D Natives, September 5, 2018. www.3dnatives.com.
16. Adrian Willings and Luke Edwards, "Best 3D Prints: The Crazy and Coolest Things People Have Printed," Pocket-Lint, January 15, 2020. www.pocket-lint.com.
17. Quoted in ICON, "New Story and ICON Unveil the First Permitted 3D-Printed Home," March 15, 2018. www.iconbuild.com.
18. Quoted in ICON, "New Story and ICON Unveil the First Permitted 3D-Printed Home."
19. Quoted in Jack Colyer, "Ashley Furniture Transforms Factory Floor with 3D Printing," 3D Printing Industry, May 31, 2019. https://3dprintingindustry.com.
20. Quoted in Katy B. Olson, "Designers Embrace 3D-Printed Furniture," Business of Home, August 9, 2017. https://businessofhome.com.
21. Quoted in Material District, "3D Printed Furniture Made from Belgian Plastic Waste," November 19, 2019. https://materialdistrict.com.
22. Charles Goulding, "Applying 3D Printing Technology to Kitchen Appliances," Fabbaloo, June 27, 2019. www.fabbaloo.com.
23. Carlota Valdivieso, "A Guide to 3D Printed Food: Revolution in the Kitchen?," 3Dnatives, February 4, 2019. www.3dnatives.com.
24. Quoted in Emma Woollacott, "Would You Eat a 'Steak' Printed by Robots?," BBC, February 25, 2020. www.bbc.com.
25. Quoted in Nancy Cates, "3D-Printed Clothing Emphasizes Innovation, Sustainability," *UV+EB Technology*, May 28, 2019. https://uvebtech.com.
26. Joseph Flynt, "50 Useful 3D Printed Objects That Will Make Your Life Easier!," 3D Insider, January 15, 2019. https://3dinsider.com.

Chapter Three: Many Marvelous Medical Uses

27. Quoted in Padraig Belton, "A New Bladder Made from My Cells Gave Me My Life Back," BBC, September 11, 2018. www.bbc.com.

28. Quoted in Belton, "A New Bladder Made from My Cells Gave Me My Life Back."
29. Quoted in Wendy Kagan, "How Far Away Are We from 3D Printing Organs?," *Chronogram*, August 1, 2019. www.chronogram.com.
30. Quoted in Brian Lord, "Brigham and Women's Hospital Successfully 3D Bioprints Blood Vessels," 3D Printing Industry, August 30, 2018. https://3dprintingindustry.com.
31. Katherine Plewa, "The Most Promising 3D Printed Organs Projects (2020 Update)," Sculpteo, October 16, 2019. www.sculpteo.com.
32. Quoted in Global Health Newswire, "Handheld 3D Skin Printer Demonstrates Accelerated Healing of Large, Severe Burns," February 4, 2020. https://globalhealthnewswire.com.
33. Quoted in Arlene Lo, "FabRx 3D Prints Personalized Tablets for Treatment of Rare Metabolic Disease in Children," 3D Printing Industry, September 12, 2019. https://3dprintingindustry.com.
34. Mark Jones, "A Personalized Dose: 3D Printed Medicine in Big Pharma," TechHQ, July 31, 2019. https://techhq.com.
35. Jones, "A Personalized Dose."
36. David H. Gorski, "Are Medical Errors Really the Third Most Common Cause of Death in the U.S.? (2019 Edition)," Science-Based Medicine, 2019. https://sciencebasedmedicine.org.
37. Farai Mashambanhaka, "Medical 3D Printing: The Best Healthcare Applications," All3DP, March 17, 2019. https://all3dp.com.

Chapter Four: New Life for Ancient Artifacts

38. Quoted in Charlotte Coates, "How Are Some of the World's Best Known Museums Doing Amazing Things with 3D Printing?," MuseumNext, December 1, 2019. www.museumnext.com.
39. Coates, "How Are Some of the World's Best Known Museums Doing Amazing Things with 3D Printing?"
40. Quoted in Coates, "How Are Some of the World's Best Known Museums Doing Amazing Things with 3D Printing?"

41. Quoted in Metropolitan Museum of Art, "About Met," Thingiverse, 2020. www.thingiverse.com.
42. Quoted in Simplify3D, "Restoring Museum Relics with 3D Printing," May 15, 2019. www.simplify3d.com.
43. Quoted in Katherine J. Wu, "Smithsonian Releases 2.8 Million Images into Public Domain," *Smithsonian*, February 25, 2020. www.smithsonianmag.com.
44. Quoted in Tyler Koslow, "Wessex Archaeology 3D Scans and Prints Two Shipwrecks from Underwater Archaeological Sites in UK," 3DPrint.com, May 24, 2016. https://3dprint.com.
45. Quoted in Morgan Hamel, "Prehistoric Preservation: 3D Printing Dinosaur Bones at SWAU," Re3D, April 2, 2018. https://re3d.org.
46. Quoted in Hamel, "Prehistoric Preservation."
47. Quoted in Hamel, "Prehistoric Preservation."
48. Quoted in Hamel, "Prehistoric Preservation."

Chapter Five: 3-D Printing in the Final Frontier

49. Bryce Salmi, "The World's Largest 3D Metal Printer Is Churning Out Rockets," *IEEE Spectrum*, October 25, 2019. https://spectrum.ieee.org.
50. Salmi, "The World's Largest 3D Metal Printer Is Churning Out Rockets."
51. Salmi, "The World's Largest 3D Metal Printer Is Churning Out Rockets."
52. NASA, "What Is Orion?," October 23, 2019. www.nasa.gov.
53. Quoted in Mae Rice, "14 3D Printing Applications & Examples," BuiltIn, December 13, 2019. https://builtin.com.
54. Quoted in Stephanie Hendrixson, "Orion Gets a Lift from Electro-Static Dissipative (ESD) Material," Additive Manufacturing, July 20, 2018. www.additivemanufacturing.media.
55. Quoted in NASA, "Space Station 3-D Printer Builds Ratchet Wrench to Complete First Phase of Operations," December 22, 2014. www.nasa.gov.
56. Quoted in ASM International, "Asteroid Metal Used to 3D-Print Logo," February 8, 2016. www.asminternational.org.

57. Quoted in ASM International, "Asteroid Metal Used to 3D-Print Logo."
58. Search and Clouds AO, "Mars Ice House: 3D Printing with Ice," Mars Ice House, 2015. www.marsicehouse.com.
59. Quoted in Billy Hurley, "3D Printing and Space Exploration: How NASA Will Use Additive Manufacturing," Tech Briefs, January 17, 2020. www.techbriefs.com.

FOR FURTHER RESEARCH

Books

Craig E. Blohm, *3D Printing and Medicine*. San Diego: ReferencePoint, 2018.

Keon Boozajomehri, *3D Printing at School and Makerspaces*. New York: Cavendish Square, 2018.

Lydia Cline, *3D Printing Projects for Makerspaces*. New York: McGraw-Hill, 2017.

Hal Marcovitz, *What Is the Future of 3D Printing?* San Diego: ReferencePoint, 2017.

Cecilia Pinto McCarthy, *How 3D Printing Will Impact Society*. San Diego: ReferencePoint, 2019.

Ben Redwood et al., *The 3D Printing Handbook*. Amsterdam: 3D Hubs, 2017.

3Doodler, *What Will You Create?* New York: Wobbleworks, 2019.

Internet Sources

Donovan Alexander, "6 Challenges of 3D Printing for Space Voyages," Interesting Engineering, July 31, 2019. https://interestingengineering.com.

Charlotte Coates, "How Are Some of the World's Best Known Museums Doing Amazing Things with 3D Printing?," MuseumNext, December 1, 2019. www.museumnext.com.

Jamie D., "The Best 3-D Printing Applications in Space," 3D natives, May 4, 2019. www.3dnatives.com.

Lucie Gaget, "Meet the Urban Transportation of the Future Made with 3D Printing," Sculpteo, April 11, 2018. www.sculpteo.com.

Aristos Georgiou, "Tiny Replicas of Michelangelo's *David* Created by Scientists Using Advanced 3D Printing Technique," *Newsweek*, December 20, 2019. www.newsweek.com.

Dana Goldberg, "History of 3D Printing: It's Older than You Are (That Is, If You're Under 30)," *Redshift* (blog), AutoDesk, April 13, 2018. www.autodesk.com.

Farai Mashambanhaka, "Medical 3D Printing: The Best Healthcare Applications," All3DP, March 17, 2019. https://all3dp.com.

Genaro Ponce, "3D Printed Clothes: Myth or Reality?," All3DP, June 27, 2019. https://all3dp.com.

Emily Price, "How to 3D Print Some of the Smithsonian's Artifacts at Home," *Lifehacker* (blog), September 5, 2019. https://lifehacker.com.

Ben Redwood, "The Definitive 3D Printing Glossary," 3D Hubs, no date. www.3dhubs.com.

Carlota Valdivieso, "University of Maine Creates the World's Largest 3D Printed Boat," 3Dnatives, October 16, 2019. www.3dnatives.com.

Emma Woollacott, "Would You Eat a 'Steak' Printed by Robots?," BBC, February 25, 2020. www.bbc.com.

Websites

3D Printers, Explain That Stuff! (www.explainthatstuff.com/how-3d-printers-work.html). This useful site created by British science writer Chris Woodford offers a thorough but readable and understandable explanation for how 3-D printing actually works. More than thirty links lead to external explanations.

3D Printing: The Complete Engineering Guide (www.3dhubs.com/guides/3d-printing/). This comprehensive site shows readers not only what 3-D printing is all about but also provides many examples of both materials that can be used for printing and objects that can be printed.

A New Spin on 3D Printing Can Produce an Object in Seconds (https://arstechnica.com/gadgets/2020/02/a-new-spin-on-3d-printing-can-produce-an-object-in-seconds/). This helpful site explains some of the newest approaches to 3-D printing that emphasize increased speed in what is normally a fairly slow process. Various links lead to separate sites that explain technical terms.

Note: Boldface page numbers indicate illustrations.

PICTURE CREDITS

Cover: HelloRF Zcool/Shutterstock.com

8: Maury Aaseng
12: mimagephotography/Shutterstock.com
15: MBR/MCT/Newscom
18: MikeDotta/Shutterstock.com
24: Stephane Mahe/Reuters/Newscom
26: Jonathan Weiss/Shutterstock.com
29: Maksym Kaharlyk/Shutterstock.com
35: AlexLMX/iStock
38: Dario Sabljak/Shutterstock.com
41: Gumpanat/Shutterstock.com
46: Heracles Kritikos/Shutterstock.com
49: Akimov Igor/Shutterstock.com
53: Ruslan Ayupov/Shutterstock.com
57: NASA
58: NASA/Marvin Smith
62: ESA 2010 MPS for OSIRIS Team MPS/UPD/LAM/IAA/RSSD/INTA/UPM/DASP/IDA

ABOUT THE AUTHOR

Classical historian, amateur astronomer, and award-winning author Don Nardo has written numerous volumes about scientific topics, including *Destined for Space* (winner of the Eugene M. Emme Award for best astronomical literature); *Tycho Brahe* (winner of the National Science Teacher's Association's best book of the year); *Planet Under Siege: Climate Change*; *Deadliest Dinosaurs*; and *The History of Science*. Nardo, who also composes and arranges orchestral music, lives with his wife, Christine, in Massachusetts.